THE WOODEN SATELLITE HANDBOOK

Transforming Raw Facts into Strategic
Knowledge

SEAN T. ROLAND

COPYRIGHT

TABLE OF CONTENTS

INTRODUCTION
Wooden Satellite Handbook

In a world where the frontiers of technology seem to know no bounds, we are constantly exploring innovative ways to tackle some of our planet's most pressing challenges. At the heart of this exploration lies a fascinating concept: wooden satellites. Yes, wood – a material historically associated with warmth, earthiness, and humble origins – is now being seen as a revolutionary asset in space technology. This new perspective invites us to look at sustainability, resourcefulness, and functionality in ways that many of us have never imagined.

The *Wooden Satellite Handbook* serves as a comprehensive guide to this emerging realm, where we blend the seemingly mundane with the high-tech to address an unexpected, yet urgent, concern: the impact of space exploration on our environment. This book isn't just about the technicalities of using wood in satellite construction; it's about understanding the motivation, science, and philosophy behind this unique approach. In a world dominated by advancements in synthetic materials and metal alloys, wooden satellites offer a counterpoint, a symbol of humanity's drive to innovate responsibly.

Why Wooden Satellites?

When you think about a satellite, the mind typically conjures images of metal structures orbiting Earth, equipped with state-of-the-art technology. However, these structures are not without their drawbacks. Traditional satellites, constructed with metals and high-grade composites, often contribute significantly to space debris once they reach the end of their operational life. This debris orbits the Earth indefinitely, posing collision risks to functioning satellites, spacecraft, and potentially, future missions. Not only does this debris threaten the success of space exploration, but it also has implications for the safety of Earth's atmosphere and orbital environments.

Wooden satellites present a pioneering solution to the problem of space debris. Wood, being organic, is a biodegradable material that burns up more efficiently upon re-entry into Earth's atmosphere. Instead of contributing to the mounting space debris problem, wooden satellites offer a more environmentally friendly alternative, one that aligns with the ethos of sustainability and conservation. With the growing global focus on eco-friendly practices, it's an approach that invites admiration, curiosity, and hope for a cleaner, safer future in space exploration.

The History and Inspiration Behind Wooden Satellites

The concept of using wood as a satellite material might seem novel today, but it has its roots in humanity's long history with this versatile resource. Wood has been used for centuries, not just as a building material, but as a symbol of adaptability, durability, and resilience. From wooden ships that sailed uncharted oceans to bridges that have withstood centuries, wood has been part of our journey into new frontiers.

In recent years, scientists and engineers began to ask the question: "Why not use wood for space exploration?" This inquiry opened doors to a realm of possibilities. As we explore the history behind this concept, we find that wooden satellites are inspired by ancient practices of innovation and resourcefulness. The idea of using wood in space not only draws from our understanding of traditional materials but also aims to extend the values of sustainability and ingenuity to the modern technological era.

A Glimpse into the Science and Engineering

Understanding how wood behaves in space is essential for appreciating the journey toward making wooden satellites a reality. While wood is strong and durable, the extreme conditions of space – temperature fluctuations, exposure to cosmic radiation, and the vacuum of space – demand rigorous testing and engineering finesse. To succeed in this endeavour, scientists are exploring different types of wood, identifying those with the best properties to withstand the hostile environment of space. They consider factors such as the wood's grain, density, and resilience against radiation. Through these explorations, specific wood types and treatments are chosen to ensure they meet the stringent demands of satellite construction.

In addition to material selection, engineers are developing new techniques for treating and assembling wooden satellite components. These advancements highlight how modern science can adapt traditional materials for futuristic applications. For instance, research has shown that specially treated wood can endure high levels of temperature and resist damage from micrometeoroids, enhancing its feasibility as a satellite material. This book delves into the science and engineering behind these techniques, offering insights into the processes that make wooden satellites possible.

Environmental Benefits and Sustainability

The environmental advantages of wooden satellites extend beyond their ability to degrade more easily in the atmosphere. By replacing some of the materials traditionally used in satellites with wood, we reduce reliance on metals and synthetic compounds that are often mined or produced through environmentally taxing processes. Using wood contributes to a reduction in the carbon footprint of satellite construction and encourages the exploration of renewable materials for future aerospace projects.

Moreover, the use of wood sends a message about the importance of sustainable innovation. In the space industry, where advanced materials are typically sought after, the decision to explore organic, biodegradable options signifies a shift towards responsible engineering. This shift reflects broader global efforts to address the environmental impact of our activities, even those beyond Earth's atmosphere. This handbook sheds light on how wooden satellites fit into the larger narrative of sustainability and why their development marks an important milestone in the aerospace industry.

Addressing Challenges and Skepticism

As with any groundbreaking idea, wooden satellites are not without their sceptics. Many question the durability, longevity, and practicality of using a natural material in space. After all, the hostile conditions outside Earth's atmosphere are far from the environments where wood has traditionally thrived. This book addresses these concerns directly, exploring the challenges of working with wood and how researchers are overcoming them.

For instance, wood's vulnerability to fire may initially seem like a limiting factor; however, the vacuum of space eliminates the need for oxygen, rendering fire hazards non-existent in orbit. Similarly, advanced treatments and coatings can protect wood from radiation and thermal stress. These scientific breakthroughs are critical to understanding how wooden satellites can perform in space as effectively as conventional satellites, and they underscore the ingenuity of this innovative approach.

Who Should Read This Handbook?

The *Wooden Satellite Handbook* is designed for a diverse audience. For engineers and scientists, it offers a deep dive into the technical aspects of constructing a satellite from wood, with chapters that detail the engineering principles, material treatments, and testing procedures that go into making this vision a reality. For environmental enthusiasts, this book provides a hopeful narrative about how sustainable materials can transform industries traditionally resistant to change. For those interested in the future of space exploration, this handbook opens a window into how we might redefine our approach to off-planet ventures.

Beyond these groups, this book will resonate with anyone who values innovation and sustainability. The story of wooden satellites is not just about technology but about finding creative solutions to complex problems – a theme that transcends industry boundaries and speaks to the potential for human ingenuity in all aspects of life.

The Road Ahead

The journey toward wooden satellites is still in its early stages, but the momentum is growing. As we look to the future, we see opportunities not just for wooden satellites, but for the use of renewable materials in other aerospace applications. This movement signals the beginning of a new era where sustainability is not an afterthought but a driving force in innovation.

The *Wooden Satellite Handbook* serves as both a guide and a beacon for what lies ahead. By understanding the science, the motivation, and the potential of wooden satellites, we are participating in a larger dialogue about our responsibilities as explorers, innovators, and stewards of our planet.

As we explore the pages of this book, let us keep an open mind and an open heart. The wooden satellite is more than just a satellite; it is a testament to the fact that even as we reach for the stars, we can remain grounded in values that honour the Earth. Let us embark on this journey together, exploring the possibilities that arise when technology meets responsibility, and when innovation respects the planet that has always been our home.

CHAPTER 1
The Idea Behind a Wooden Satellite

Space exploration is one of humanity's boldest ventures, requiring constant innovation to reach new frontiers. Among the latest concepts attracting attention is the idea of a wooden satellite—a project that, at first glance, seems unconventional, even surprising. Why would scientists and engineers consider using wood, a material with ancient roots and associations with nature, in the high-tech environment of space? The journey to this idea is rooted in both creativity and a modern urgency to find sustainable, environmentally-friendly solutions to pressing challenges. This chapter will explore the origins of the concept, the significant roles of Kyoto University and Sumitomo Forestry in bringing this idea to life, and the potential benefits that wooden satellites could introduce to space exploration. Let's dive into each of these fascinating aspects, starting with the initial spark that led to this unlikely yet groundbreaking concept.

Origins of the Concept

The idea of using wood in a satellite may seem far-fetched at first. Space environments are harsh, demanding materials that can withstand extreme conditions, from drastic temperature shifts to cosmic radiation and the vacuum of space. Traditional satellite construction relies on metals and composites that can handle these stresses. However, with thousands of satellites orbiting Earth and more launched each year, a growing problem has become apparent: space debris.

Space debris, also known as orbital debris, consists of defunct satellites, abandoned launch stages, and fragments resulting from collisions and disintegration. These objects remain in Earth's orbit indefinitely, posing threats to operational satellites and future space missions. As these fragments accumulate, they increase the risk of collisions, which could render certain orbits unusable. Addressing this issue has led researchers to consider materials that are more biodegradable upon re-entry into the Earth's atmosphere.

Wood, an organic material, is not only biodegradable but also has properties that might suit it for certain applications in space. If a satellite made of wood were to re-enter Earth's atmosphere, it would burn up cleanly, leaving minimal or no debris in orbit. This would effectively address some of the space debris issues associated with traditional satellite materials. By considering wood, researchers are challenging traditional assumptions in space engineering, pushing the boundaries of what we perceive as viable materials for space use.

This exploration is not simply about aesthetics or novelty; it reflects a deeper awareness of the need for sustainability in every field, including space exploration. The wooden satellite concept was born out of a combination of environmental responsibility and curiosity, as scientists asked themselves, "Can we make satellites that are as kind to space as they are effective?"

The Role of Kyoto University and Sumitomo Forestry

The concept of wooden satellites gained traction through the collaborative efforts of Kyoto University and Sumitomo Forestry, a leading Japanese forestry and wood products company. Together, they represent a fusion of academic expertise and industrial experience, aiming to advance this unique idea from theoretical exploration to practical implementation.

Kyoto University, known for its pioneering research and commitment to sustainability, has been involved in various initiatives to promote environmentally-friendly technologies. Scientists and researchers at Kyoto have long been interested in sustainable solutions that address global challenges, from alternative energy sources to eco-friendly building materials. With the rise of the space debris problem, the university saw an opportunity to apply its environmental focus to the realm of aerospace.

Sumitomo Forestry, on the other hand, brings a wealth of experience in wood technology, forestry management, and material science. With centuries-old expertise in understanding wood's structural, aesthetic, and functional properties, Sumitomo Forestry provides the technical know-how to make wood suitable for environments beyond its traditional uses. The company's commitment to sustainability aligns well with Kyoto University's goals, making the collaboration a natural fit.

In 2020, Kyoto University and Sumitomo Forestry announced a joint project to develop a wooden satellite, aiming to launch the world's first satellite of this kind by 2023. The project's objectives were clear: identify types of wood that could endure the conditions of space, develop treatments to enhance wood's durability, and design a satellite structure that would be efficient and operational in orbit. The project symbolised a commitment to pushing scientific boundaries while remaining grounded in environmentally-responsible principles.

Through their partnership, Kyoto University and Sumitomo Forestry conducted a series of studies to assess how different types of wood responded to space-like conditions. Experiments were performed to simulate vacuum, thermal, and radiation exposures, allowing researchers to determine the most suitable wood varieties and treatments for satellite construction. This partnership demonstrated that innovation does not always require new, synthetic materials; sometimes, it involves revisiting and reimagining traditional resources in entirely new contexts.

Potential Advantages of Wood in Space

The choice of wood as a material for satellite construction introduces a range of unique benefits that may have substantial implications for the future of space technology. Although it is an unconventional choice, wood offers distinct advantages that could make it a valuable asset in specific applications.

1. **Environmental Sustainability and Degradability**
   The primary benefit of wooden satellites is their potential to reduce space debris. Wood, unlike metals and synthetic composites, is biodegradable. When a wooden satellite re-enters Earth's atmosphere at the end of its operational life, it will burn up cleanly,

leaving minimal traces behind. This property could significantly reduce the amount of long-lasting debris in Earth's orbit. In contrast, traditional satellites often contribute to the accumulation of space debris, which can remain in orbit for decades or even centuries.

By opting for a material that can fully degrade, scientists and engineers are taking a proactive approach to space sustainability. This aligns with global efforts to limit the environmental impact of technological advancements. In the context of space exploration, where sustainability is often an afterthought, wooden satellites represent a meaningful step toward environmentally-conscious engineering.

2. **Thermal Stability and Insulation Properties**
   Wood has natural insulating properties, which can be beneficial in space environments where temperatures fluctuate drastically. Unlike metals that conduct heat and cold quickly, wood's cellular structure allows it to insulate more effectively. This means that wooden components might be less affected by extreme temperatures, helping to protect the satellite's internal instruments from sudden thermal changes.

Although wood is not commonly associated with high-tech applications, its thermal stability and insulation properties have been well-studied in fields like construction and architecture. By adapting wood for use in satellites, researchers are leveraging its insulating capabilities to potentially create satellites that require less energy for temperature regulation. This could translate into greater efficiency and lower power consumption, which are important factors for long-term space missions.

3. **Lightweight and Strong**

    Despite being relatively lightweight, wood can be remarkably strong, depending on the type and treatment used. This strength-to-weight ratio is advantageous for satellite construction, where every kilogram matters. Using wood as a structural component could potentially reduce the overall weight of a satellite, making it easier and more cost-effective to launch. By reducing launch costs, wooden satellites may allow more organisations to participate in space exploration, democratising access to space-based research and technology.

Furthermore, the unique grain and cellular structure of wood allow it to absorb and distribute stress differently than metals or composites, potentially providing resilience against micro-impacts from space debris. Although metals are conventionally chosen for their rigidity and durability, wood's strength, when treated correctly, could hold its own in space's demanding environment.

4. **Low Electromagnetic Interference**

In the realm of satellite technology, electromagnetic interference (EMI) is a constant concern. EMI can disrupt the satellite's instruments and communication signals, causing data inaccuracies or even failure of the satellite's systems. Metals, commonly used in satellite construction, can contribute to EMI due to their conductive properties. Wood, being a natural insulator, does not conduct electricity and therefore generates minimal electromagnetic interference. By constructing satellites with wooden components, engineers could mitigate issues related to EMI, potentially improving the reliability of onboard instruments and communications systems.

5. **Symbolic and Educational Value**

Beyond its technical advantages, the concept of a wooden satellite holds symbolic significance. It challenges traditional perceptions of technology and innovation, reminding us that the future does not have to be built solely on synthetic or rare materials. The wooden satellite serves as a symbol of humanity's commitment to balancing progress with sustainability, fostering a mindset that values environmentally-conscious development.

Additionally, the project has educational value. By exploring alternative materials like wood, scientists are inspiring the next generation of engineers, researchers, and environmentalists to think outside the box. The wooden satellite can serve as a teaching tool, demonstrating that creativity, responsibility, and scientific rigour can go hand-in-hand.

Bringing the Concept to Reality: Challenges and Considerations

While the idea of a wooden satellite is captivating, its realisation requires addressing several challenges. Wood, as a natural material, does not inherently possess the qualities necessary to withstand the hostile conditions of space. To make wooden satellites feasible, researchers must carefully consider how to preserve wood's beneficial properties while enhancing its resilience.

Selecting the Right Type of Wood

Different types of wood have different properties, including density, grain pattern, and resistance to environmental stresses. Researchers must identify which types of wood can endure space's unique conditions. Hardwoods, known for their durability, are likely candidates, but each variety has unique strengths and weaknesses. For example, some types of wood may be more resistant to thermal stress, while others may offer better strength-to-weight ratios. Kyoto University and Sumitomo Forestry are exploring various options, testing them under simulated space conditions to determine the most suitable type for satellite construction.

Treatment and Coating Techniques

To protect wood from the radiation, vacuum, and extreme temperatures of space, it must undergo specific treatments and coatings. These treatments may involve chemical processes that make the wood more resistant to moisture, prevent warping or cracking, and shield it from UV and cosmic radiation. Sumitomo Forestry's expertise in wood treatment is crucial here, as their innovative coatings and treatments help prepare wood for these challenging conditions without compromising its structural integrity.

Designing for Functionality and Efficiency

Designing a satellite from wood involves rethinking structural components. Engineers must carefully consider how to integrate wooden elements with other materials like metals and plastics, balancing the advantages of each to create a satellite that is both functional and efficient. This requires a multidisciplinary approach, combining insights from aerospace engineering, material science, and forestry.

Testing in Real-Life Conditions

Testing is a crucial phase of the wooden satellite project. While simulations and laboratory tests provide valuable insights, real-life testing is essential to confirm the material's behaviour in space. The project team is planning to conduct tests in near-space environments to gather data on how wooden satellites perform in orbit. These tests will be instrumental in refining designs, improving treatments, and optimising functionality for the final launch.

A Pioneering Step Toward Sustainable Space Exploration

The wooden satellite project, spearheaded by Kyoto University and Sumitomo Forestry, represents a forward-thinking approach to space exploration. By reconsidering materials traditionally used in satellite construction, these researchers and engineers are not only tackling practical issues like space debris but also promoting a more responsible form of technological innovation.

The concept of a wooden satellite challenges the status quo, inviting scientists, environmentalists, and the public alike to reimagine what is possible when sustainability becomes a central focus of technological development. It serves as a reminder that materials we often overlook—like wood—may hold the key to unlocking more sustainable solutions in unexpected places.

As we proceed with this book, we will explore the technical challenges, scientific breakthroughs, and the inspiring mindset behind the wooden satellite project. This journey is as much about engineering as it is about creativity, courage, and the belief that space exploration can indeed be compatible with the values of environmental stewardship and sustainable progress.

CHAPTER 2
Materials and Design of LignoSat

The prospect of a wooden satellite seems almost paradoxical in a field dominated by high-tech alloys and advanced composites. Yet, as we embrace sustainable and innovative approaches in every aspect of life, even space exploration is experiencing a shift. The LignoSat project, a wooden satellite spearheaded by Kyoto University and Sumitomo Forestry, represents a bold step toward reducing the environmental footprint of space missions. This chapter explores the careful selection of materials for LignoSat, especially the choice of magnolia wood, the engineering hurdles that the team faced, and the wider implications of eco-friendly spacecraft materials. This chapter aims to provide a deep understanding of how traditional materials like wood are ingeniously adapted to meet the rigorous demands of outer space, paving the way for a more sustainable future in satellite technology.

Choosing the Right Wood: Why Magnolia?

Choosing the right wood for LignoSat was not a decision made lightly. In fact, it was the result of extensive research and experimentation to identify a wood that would meet the unique demands of space. Out of all the possible types of wood, magnolia emerged as the preferred choice due to its combination of durability, flexibility, and structural integrity.

Magnolia, a hardwood native to both East Asia and the Americas, was chosen for its remarkable properties. While it may not have the immediate visual appeal or strength of traditional metals, magnolia possesses a unique balance of characteristics that make it particularly suitable for space applications. Below are some of the key reasons why magnolia wood was selected for LignoSat:

1. **Density and Strength**

   Magnolia wood has a moderate density and high strength-to-weight ratio. It is not as dense as some hardwoods like oak or mahogany, which makes it light enough for satellite construction, where every gram counts. However, it still possesses enough strength to withstand the stress of launch and the harsh conditions of space. The balance between strength and weight makes magnolia an ideal candidate, as it reduces the satellite's overall mass without compromising durability.

2. Thermal Insulation Properties

Wood is an organic material with inherent insulating properties. In space, where temperatures fluctuate dramatically between extreme heat and cold, magnolia wood's natural insulation can help protect the satellite's internal components. Unlike metals, which conduct heat readily, magnolia wood can provide a more stable thermal environment for the satellite's instruments. This thermal stability is essential for ensuring the satellite's functionality and prolonging its operational lifespan.

3. Grain and Microstructure

The grain structure of magnolia wood is more uniform compared to many other types of wood. This uniformity provides a predictable behavior under stress, which is crucial for space applications where consistency and reliability are paramount. Magnolia's microstructure also makes it more resistant to splitting or cracking under pressure, further increasing its resilience in the vacuum and microgravity of space.

4. **Ease of Treatment and Modification**

Magnolia wood can undergo various treatments to enhance its durability and resistance to radiation and other space conditions. Researchers can treat the wood to increase its resistance to moisture, fungi, and UV radiation, which are essential factors for materials in space. Magnolia is also relatively easy to shape and modify, allowing for precision engineering and adjustments during the construction of LignoSat.

5. **Environmental Sustainability**

In addition to its technical properties, magnolia wood aligns with the project's environmental goals. As a renewable resource, it offers a much smaller carbon footprint than traditional metals. By choosing wood, the LignoSat project exemplifies a commitment to reducing environmental impact, demonstrating that even in space exploration, sustainability can be prioritized.

Engineering Challenges: Designing a Wooden Satellite for Space

Although wood's unique properties make it a viable option for satellite construction, designing LignoSat presented a variety of engineering challenges. The extreme environment of space imposes stresses and conditions far beyond what most materials encounter on Earth. From radiation to vacuum exposure and temperature fluctuations, each aspect required careful planning and innovation. Below are some of the major challenges faced in engineering a wooden satellite that could withstand space conditions.

1. **Resistance to Cosmic Radiation**

   Space is filled with high-energy radiation, which can degrade materials over time. Traditional satellite materials, such as aluminum and titanium, are chosen partly for their resistance to this radiation. Wood, however, is an organic material that is vulnerable to degradation when exposed to cosmic radiation. To address this issue, researchers applied specialized coatings to magnolia wood to enhance its resistance to radiation. These coatings help protect the wood's cellular structure, ensuring that it can withstand prolonged exposure to space's radiation environment without significant degradation.

2. **Structural Integrity in Microgravity**

   The microgravity conditions of space can have unique effects on the structural stability of materials.

While wood is strong, it has a certain level of flexibility that could be problematic in space if not managed properly. Engineers had to account for this flexibility to ensure that the wooden components would not warp or deform over time. Through testing and modelling, they were able to design a structure that maintains its shape and functionality in a low-gravity environment.

3. **Thermal Stability in Extreme Temperatures**

The temperature in space can range from extremely hot when exposed to the sun to freezing cold in the shadow of Earth or other celestial bodies. Wood, by nature, is not as thermally conductive as metals, which in some ways is advantageous for maintaining a stable internal temperature. However, extreme temperatures can still cause wood to expand or contract, potentially compromising structural integrity. To mitigate this, researchers applied thermal-resistant coatings and treatments to the magnolia wood, allowing it to better handle temperature fluctuations without warping or cracking.

4. Protection from Micrometeoroids

Space is filled with tiny particles, including micrometeoroids, that travel at incredibly high speeds. Even small particles can cause damage when they collide with a satellite. Traditional satellite materials are often chosen for their ability to withstand these impacts. For LignoSat, engineers developed a protective outer layer that shields the wood from micrometeoroid impacts. This layer is essential for preserving the wood's structure and preventing punctures or fractures that could compromise the satellite's functionality.

5. Vacuum Exposure and Outgassing

In the vacuum of space, materials can undergo a process known as outgassing, where trapped gases are released. This phenomenon can cause materials to deteriorate and lead to a buildup of gases within the satellite. Since wood is an organic material with a porous structure, it is prone to outgassing. Engineers needed to seal the magnolia wood to prevent outgassing and ensure that the satellite's performance would not be compromised. Through vacuum treatments and sealing techniques, they minimized outgassing while preserving the wood's structural integrity.

6. **Ensuring Durability and Longevity**

Satellites are expected to remain operational for several years, requiring materials that are not only durable but also resistant to long-term wear. Engineers performed extensive durability tests on magnolia wood, subjecting it to prolonged exposure to simulated space conditions. These tests helped researchers identify potential points of failure and refine the satellite's design to enhance its durability. The goal was to ensure that LignoSat could remain functional throughout its mission, even in the face of space's harsh conditions.

Innovation in Sustainable Spacecraft Materials

The LignoSat project has drawn attention not only for its unique use of wood but also for its role in advancing eco-friendly materials for spacecraft. While wood is the primary focus of LignoSat, it has sparked broader interest in sustainable materials for space exploration. Researchers and engineers are exploring alternatives to traditional materials that can meet the demands of space without contributing to environmental degradation. This section introduces some of the eco-friendly materials being considered for future satellite projects and highlights the innovative approaches being taken to make space technology more sustainable.

1. **Biodegradable Polymers**

   Biodegradable polymers, derived from natural sources like cornstarch or cellulose, offer an alternative to synthetic plastics. These materials can degrade more easily upon re-entry, reducing space debris. Although not as widely used in aerospace applications yet, research is ongoing to make biodegradable polymers suitable for satellite components, particularly for structural elements and housing.

2. **Recycled and Recyclable Materials**

   The concept of a circular economy, where materials are reused and recycled, is gaining traction in the space industry. Some companies are exploring the use of recycled metals and composites in satellite construction, reducing the need for raw resources. Additionally, designing satellites with recyclable components allows for easier recovery and reuse at the end of their operational lives, contributing to a more sustainable lifecycle.

3. Nano-enhanced Materials

Nanotechnology is providing opportunities to enhance the properties of eco-friendly materials. By adding nanoparticles to wood or biodegradable polymers, engineers can improve their strength, thermal stability, and resistance to radiation. For example, adding carbon nanotubes to biodegradable polymers can increase their structural integrity, making them viable for use in space.

4. Self-Healing Materials

Self-healing materials are designed to repair small damages on their own, which could extend the lifespan of satellites and reduce maintenance needs. Although still in the research phase, self-healing materials are being considered for future spacecraft applications. They hold particular promise for protecting satellites from micrometeoroid impacts, as the material could automatically seal minor punctures or cracks.

5. Organic-Based Insulators

Organic-based insulation materials, like aerogels derived from natural sources, offer lightweight and effective thermal protection. These materials can be used alongside wood in eco-friendly satellites, providing additional insulation without contributing to the environmental impact. Organic-based insulators are particularly beneficial for satellites, as they reduce reliance on synthetic insulation materials that can be difficult to dispose of safely.

6. **Solar-Powered Systems with Green Batteries**
   Power systems are a significant component of satellites, and traditional power sources often rely on non-renewable materials. Solar power is a common choice for satellites, but green battery technology, such as lithium-ion batteries derived from recycled sources, is gaining popularity. These green batteries can be combined with solar panels to create power systems that align with sustainable practices. By integrating green power systems, eco-friendly satellites like LignoSat can further reduce their environmental impact.

Pioneering a Sustainable Future in Space Exploration

The development of LignoSat and its innovative use of wood highlight the potential for more sustainable practices in space exploration. As we continue to explore eco-friendly materials and design methods, the vision of a sustainable space industry becomes more attainable. Projects like LignoSat are paving the way for a new era of satellite technology that prioritizes environmental responsibility without compromising functionality.

The selection of magnolia wood, the creative engineering solutions, and the exploration of new materials demonstrate the potential to rethink traditional approaches to space missions. As we face challenges such as space debris and resource depletion, embracing sustainability in satellite design becomes increasingly critical. LignoSat is a pioneering step in this direction, proving that even in the high-tech world of space exploration, we can find room for traditional materials and innovative, eco-friendly solutions.

In the following chapters, we will delve deeper into the processes, treatments, and real-world applications of wooden satellites, offering insights into how LignoSat and similar projects could redefine our approach to space exploration and inspire future generations to pursue sustainability in all technological advancements.

CHAPTER 3

Preparing for Launch: Testing and Logistics

The journey to launching LignoSat, the world's first wooden satellite, has been one of meticulous preparation and extensive testing, pushing the boundaries of modern engineering to bring a seemingly unconventional material into the high-tech arena of space exploration. While the use of wood in satellite construction represents a bold step forward in sustainable space technology, the path to proving its viability has been filled with rigorous testing and logistical planning. To ensure that magnolia wood, the primary material used in LignoSat, could survive and operate in the harsh conditions of space, researchers conducted a series of ground-based tests to assess its resilience. This chapter delves into those critical ground-based tests, the role of key partnerships with space agencies such as SpaceX and the International Space Station (ISS), and the unique logistical challenges associated with launching a wooden satellite into orbit.

Ground-Based Tests: Proving Wood's Suitability for Space

Before even considering the logistics of launching LignoSat, researchers needed to demonstrate that magnolia wood could withstand the extreme environment of space. The challenges of space include intense cosmic radiation, wide temperature fluctuations, a vacuum with no atmospheric pressure, and exposure to micrometeoroids. Each of these factors presents a potential threat to materials, and wood, being an organic substance, required particularly rigorous testing.

To address these risks, scientists and engineers designed a comprehensive testing program for LignoSat, subjecting magnolia wood to conditions that closely mimic those it would encounter in orbit. These tests were conducted in highly specialized labs equipped to simulate space conditions as accurately as possible.

Thermal Testing: Simulating Extreme Temperatures

One of the first challenges was to determine how wood would respond to the drastic temperature changes in space. In low-Earth orbit, temperatures can swing from extremely cold when in Earth's shadow (around -250°F) to incredibly hot when exposed to the sun (up to 250°F). These fluctuations occur in cycles as the satellite orbits the planet, moving between sunlight and shadow roughly every 90 minutes.

To test wood's thermal resilience, researchers placed samples of magnolia wood in a thermal vacuum chamber, where they subjected it to rapid temperature changes similar to what it would experience in orbit. This chamber allowed them to control the temperature precisely and create a vacuum that mimicked space conditions. The goal was to observe how the wood reacted to these extremes, identifying potential issues such as cracking, warping, or any chemical breakdowns.

The results were promising. With certain treatments, including heat-resistant coatings and sealants, magnolia wood was able to withstand these temperature changes without significant deformation or damage. The wood's natural thermal stability, combined with the treatments, provided sufficient protection, allowing it to maintain structural integrity despite the harsh conditions. This test demonstrated that magnolia wood, under controlled treatment, could handle the intense thermal fluctuations of space.

Vacuum Testing: Enduring the Pressure-Free Environment

In addition to temperature, the lack of atmospheric pressure in space presents another significant challenge. In a vacuum, materials can behave differently than they would on Earth. For instance, outgassing is a common issue for organic materials in a vacuum; it refers to the release of gases that are trapped within the material. Outgassing can not only weaken the material but also contaminate sensitive instruments on the satellite.

To assess outgassing, researchers placed the wood in a vacuum chamber, simulating the pressure-free environment of space. The purpose was to observe any signs of outgassing and determine if the wood would lose mass or structural integrity. The team used spectrometry to measure the levels of gas released from the wood during the test, carefully monitoring for any components that could pose a risk to other satellite systems.

Fortunately, the vacuum treatments and specialized sealants applied to the wood proved effective in minimizing outgassing. By treating the wood beforehand, engineers reduced the likelihood of gas release, ensuring that the wood would remain stable in space. This phase of testing provided essential insights into how wood could maintain its integrity in the vacuum of space, marking another successful milestone in the satellite's development.

Radiation Exposure Testing: Protecting Against Cosmic Rays

Cosmic radiation poses a significant risk to all space materials, especially organic ones like wood. High-energy particles from the sun and distant stars can penetrate materials, causing gradual degradation over time. This radiation can weaken structural integrity and lead to failure in space, making radiation resistance an essential quality for any satellite material.

To simulate the exposure to cosmic radiation, researchers subjected magnolia wood samples to high doses of gamma radiation, which closely mimics the type of radiation the satellite would encounter in orbit. Using facilities equipped with particle accelerators, scientists bombarded the wood with radiation to assess how it would react over extended periods.

This testing showed that with the addition of certain protective coatings, magnolia wood could endure radiation levels expected in low-Earth orbit without significant deterioration. While the wood would inevitably degrade over long periods, its lifespan was sufficient for the mission's intended duration. This breakthrough demonstrated that magnolia wood, with appropriate treatment, could survive the radiation environment in space, bringing LignoSat one step closer to reality.

Impact Testing: Safeguarding Against Micrometeoroids

Space is filled with small particles, such as micrometeoroids, which travel at high speeds and can damage satellite surfaces. Traditional satellite materials, such as aluminium and titanium, are chosen partly for their ability to withstand these impacts. For LignoSat, the team had to ensure that wood would also have some level of resistance to these particles.

To test this, researchers used impact testing machines to simulate the collisions that could occur with micrometeoroids. By firing small projectiles at high velocities into magnolia wood samples, they measured the material's ability to withstand these impacts without cracking or breaking. Although wood is not as resistant to impacts as metals, the results showed that magnolia wood, when paired with an outer protective layer, could survive minor impacts without compromising the satellite's structure.

Collaborations with Space Agencies: Partnering with SpaceX and the ISS

The journey to launch LignoSat involved not only technical advancements but also significant collaborations with space agencies and industry leaders. Key among these partnerships were the collaborations with SpaceX and the International Space Station (ISS), both of which played crucial roles in bringing the wooden satellite project to fruition.

SpaceX: A Launch Partner for LignoSat

SpaceX, a leader in commercial spaceflight, was instrumental in providing launch services for LignoSat. Their reusable Falcon 9 rocket, known for its cost-effective and reliable launches, was an ideal choice for deploying the wooden satellite into orbit. SpaceX's reputation for innovation and commitment to advancing sustainable space exploration made them a fitting partner for LignoSat, aligning well with the project's environmental goals.

The partnership with SpaceX provided LignoSat with a secure and timely launch window. By collaborating with SpaceX, the LignoSat team could plan and coordinate their launch preparations effectively, ensuring that all technical requirements for payload integration were met. SpaceX's extensive experience in payload deployment and management allowed the team to focus on refining the satellite's design and testing protocols, knowing that the actual launch would be handled by one of the most capable players in the industry.

Furthermore, working with SpaceX provided an opportunity for LignoSat to reach a larger audience. The launch of the first wooden satellite garnered significant media attention, sparking public interest and awareness about sustainable space exploration. SpaceX's platform amplified the project's visibility, helping to educate the public about the possibilities of eco-friendly materials in space.

The International Space Station (ISS): Testing in Real Orbit Conditions

The International Space Station, operated by NASA, Roscosmos, and international partners, served as a critical testing ground for LignoSat before its final deployment into orbit. The ISS provides a unique environment where researchers can test materials and technology in actual space conditions while maintaining the ability to monitor and adjust experiments in real time.

Before launching LignoSat independently, the team arranged for samples of magnolia wood to be tested on the ISS. This allowed researchers to observe how the wood behaved under true space conditions, gathering data on radiation exposure, thermal stability, and microgravity effects. The ISS tests were invaluable, as they provided insights that ground-based simulations could not fully replicate.

By conducting experiments on the ISS, the team could refine treatments and adjust designs based on real-world data. The ISS partnership thus bridged the gap between theoretical testing and practical application, ensuring that LignoSat was fully prepared for its independent journey in orbit.

Logistical Challenges of Launching a Wooden Satellite

Preparing to launch a wooden satellite like LignoSat involved unique logistical considerations, as the nature of wood required adjustments that conventional satellites did not need. From transportation to payload integration, each step demanded careful planning to accommodate the specific properties and requirements of magnolia wood.

Transporting and Handling the Wooden Satellite

One of the initial logistical challenges was transporting LignoSat to the launch site. Traditional satellites are built with durable metals that can handle the rigors of transportation without much risk of damage. Wood, however, is more sensitive to changes in humidity and pressure. As a result, LignoSat required specialized containers that maintained a stable, controlled environment during transport to prevent any warping or cracking.

The transport team also took extra precautions to avoid vibrations, which could stress the wood or compromise its structural integrity. By using shock-absorbing packing materials and securing the satellite carefully, the team ensured that LignoSat arrived at the launch site in perfect condition, ready for final integration and testing.

Payload Integration and Adaptation for a Wooden Structure

Integrating LignoSat into the Falcon 9 rocket presented additional challenges, as the payload compartment was originally designed for satellites constructed from metal. Engineers had to account for the differences in density, strength, and mounting requirements when securing LignoSat within the payload bay.

To address these issues, the team developed custom mounting brackets and shock absorbers specifically designed to support LignoSat's structure. These adaptations ensured that the satellite would remain stable during the intense vibrations of launch and that the wood would not experience any undue strain. By tailoring the payload integration process, the team mitigated the risks associated with launching an unconventional material into space.

Protecting Against Environmental Conditions at the Launch Site

The launch site presented yet another set of environmental challenges, especially concerning humidity. Since wood is highly sensitive to moisture, any exposure to humid conditions could compromise its structure. To protect LignoSat, the team kept it in a climate-controlled environment up until the final stages of launch preparation.

As the countdown approached, engineers moved LignoSat from its controlled environment to the rocket's payload bay. This required precise coordination to minimize exposure to the outside atmosphere, ensuring that LignoSat remained stable and undamaged. The extra precautions paid off, as LignoSat was successfully secured in the rocket, ready for its journey to space.

A Milestone in Sustainable Space Exploration

The preparation and launch of LignoSat marked a pioneering achievement in sustainable space technology. From rigorous ground-based testing to valuable collaborations with SpaceX and the ISS, every aspect of LignoSat's journey represented a commitment to pushing the boundaries of what's possible in space exploration. Through careful planning, innovative engineering, and international partnerships, the team overcame the unique challenges posed by launching a wooden satellite.

LignoSat's successful deployment signifies a new chapter in the exploration of eco-friendly materials for space, demonstrating that sustainability and technological advancement can go hand-in-hand. This chapter serves as a testament to the power of collaboration and creativity in addressing global challenges, offering a glimpse of a future where space exploration is not only bold and groundbreaking but also deeply considerate of its impact on our planet.

CHAPTER 4

The Journey to the International Space Station

The journey of LignoSat, the world's first wooden satellite, from conception to orbit was marked by scientific innovation, meticulous engineering, and extensive collaboration. As the satellite embarked on its voyage, it carried the hopes of redefining sustainable space exploration. LignoSat's successful launch and subsequent arrival at the International Space Station (ISS) was a milestone, bringing together researchers, engineers, and space agencies with a shared vision of blending traditional materials with advanced technology for an eco-friendly future in space exploration. This chapter examines the excitement of the launch event with SpaceX, the initial stages of LignoSat's arrival and setup at the ISS, and the strategic significance of the satellite's orbital position, which allows for valuable and comprehensive data collection on its performance in space.

The Launch Event: LignoSat's Journey Begins with SpaceX

The launch of LignoSat was a historic moment, not only as a testament to the possibilities of sustainable materials in space but also as a celebration of international collaboration and technological ingenuity. LignoSat's journey to space began at the SpaceX launch site, where it was integrated into the payload compartment of a Falcon 9 rocket. The launch event attracted significant attention, with scientists, engineers, and media representatives eager to witness the first wooden satellite embark on its mission.

Preparation and Final Checks

In the days leading up to the launch, LignoSat underwent a series of final checks to ensure its readiness for space. Engineers carefully examined the satellite, verifying that all components were functioning correctly and that the wood remained structurally sound. The integration of LignoSat into the Falcon 9 rocket was a meticulous process, as the satellite's wooden structure required special considerations compared to traditional metal-based satellites. Custom mounting systems and shock absorbers were used to stabilize LignoSat, protecting it from vibrations and potential damage during launch.

The countdown to launch was marked by anticipation and excitement. As engineers completed their final checks, they reflected on the groundbreaking nature of the project. LignoSat was more than just a satellite; it represented the first step in exploring the use of biodegradable materials in space, opening the door for future sustainable designs in the aerospace industry. The launch event was attended by key representatives from Kyoto University, Sumitomo Forestry, and SpaceX, each of whom played a critical role in making the mission a reality.

Liftoff: The Falcon 9 Rocket Takes Flight

On launch day, all eyes were on the Falcon 9 rocket as it prepared to carry LignoSat into orbit. The rocket, known for its reliability and efficiency, had been chosen for its capacity to deploy a wide range of payloads, including unique and experimental satellites like LignoSat. As the countdown reached zero, the Falcon 9 ignited its engines, lifting off from the launch pad with a roar. The sight of the rocket ascending into the sky was awe-inspiring, symbolizing humanity's endless quest for exploration and innovation.

For the engineers and scientists involved in the project, the liftoff marked the culmination of years of research, development, and testing. Watching LignoSat leave Earth's surface was a moment of pride and accomplishment, underscoring the dedication of the entire team. As the Falcon 9 rocket continued its ascent, it separated from its first stage, which returned to Earth in a controlled descent, demonstrating SpaceX's commitment to reusability and sustainability—values that resonated with the LignoSat mission.

With the second stage of the rocket carrying LignoSat and other payloads into orbit, the satellite's journey to the ISS was well underway. The launch was smooth and without complications, a testament to the rigorous planning and precision of both SpaceX and the LignoSat team. Following a series of trajectory adjustments and communications with mission control, the Falcon 9's second stage approached the ISS, preparing for the next phase of LignoSat's journey.

Arrival at the ISS: LignoSat's Initial Setup for Testing

After reaching the vicinity of the ISS, LignoSat was prepared for transfer to the space station. The ISS, orbiting approximately 250 miles above Earth, provides a unique platform for testing and conducting experiments in microgravity. As LignoSat approached the ISS, the station's robotic arm, Canadarm2, was used to secure and position the satellite for initial setup and testing. The collaboration between the satellite team and ISS crew members was essential in ensuring a successful setup, allowing the satellite to begin its testing phase.

Initial Inspection and Setup by ISS Crew

Once safely on board, ISS crew members conducted an initial inspection of LignoSat. This inspection was crucial for confirming that the satellite had not sustained any damage during its journey from Earth to orbit. The team checked each of LignoSat's components, ensuring that the wood and other materials were intact and that all systems were operational. The inspection verified that the magnolia wood remained stable, showing no signs of cracking or warping from the transition into microgravity.

Following the inspection, the ISS crew began the setup process, which included activating LignoSat's systems and calibrating its instruments. This setup was vital for collecting accurate data on the wood's behavior and performance in space. The satellite was equipped with sensors to monitor radiation exposure, thermal fluctuations, and structural stability, allowing researchers to gather real-time data on how magnolia wood responds to the unique conditions of space.

The initial setup also involved configuring LignoSat's communication system, which allowed it to transmit data back to Earth. The satellite's communication system was tested to ensure that data could be sent reliably to researchers monitoring the mission. This connection was essential for real-time data collection, enabling the team on Earth to analyze LignoSat's performance continuously.

Commencement of Testing Procedures

With the initial setup complete, LignoSat was ready to begin its testing procedures. These procedures were designed to evaluate the satellite's resilience to cosmic radiation, temperature fluctuations, and vacuum conditions. By positioning LignoSat outside the ISS, the team could simulate the conditions the satellite would experience if it were independently orbiting Earth.

The testing procedures were carefully planned to maximize the insights gained from LignoSat's presence on the ISS. Data was collected over a series of experiments, with each test focusing on a specific aspect of the wood's behavior in space. The team monitored parameters such as outgassing, structural stability, and thermal conductivity, all of which are crucial for assessing wood's suitability as a material for future space missions.

The Significance of LignoSat's Orbital Position: Data Collection and Analysis

Positioning LignoSat at the ISS provided a unique opportunity for researchers to conduct in-depth studies of the satellite's performance. The ISS orbits Earth at a relatively low altitude, providing an ideal environment for experiments on the effects of radiation, microgravity, and atmospheric conditions on materials. The satellite's position allowed for real-time monitoring and data collection, providing valuable insights that could not be replicated through ground-based simulations alone.

Understanding Radiation Exposure at Low Earth Orbit

One of the primary goals of positioning LignoSat at the ISS was to study the effects of cosmic radiation on wood in low Earth orbit (LEO). The ISS's orbit exposes it to radiation levels higher than those on Earth's surface but lower than what satellites would experience in deeper space. This radiation exposure is sufficient to test the wood's resilience without subjecting it to the harsher conditions of higher orbits.

The ISS's orbit enabled researchers to observe how magnolia wood responded to radiation over an extended period. By monitoring changes in the wood's structure and durability, the team could evaluate whether it could withstand the radiation expected in LEO missions. The data collected was critical in determining wood's potential as a material for sustainable satellite construction, as it provided a baseline for understanding how biodegradable materials behave under constant radiation exposure.

Temperature Fluctuations and Thermal Stability

The ISS experiences temperature fluctuations similar to those in free space, with extreme heat on the sunlit side and freezing cold in the shadowed side. These rapid changes pose a significant challenge for satellite materials, as they can lead to thermal stress and potential degradation. LignoSat's positioning allowed researchers to monitor its response to these temperature swings, assessing whether the wood could handle the stress without warping or cracking.

Data on LignoSat's thermal stability was collected using onboard sensors that tracked temperature changes and the material's reaction to them. This information was essential for evaluating the feasibility of wood as a material for satellite construction, as it helped researchers understand how to enhance wood's thermal resilience through coatings and treatments.

Microgravity and Material Behavior

Microgravity is a unique aspect of the ISS's orbit, and studying LignoSat in this environment provided insights into how wood behaves when subjected to long-term weightlessness. Wood's microstructure, which includes cellular arrangements that can vary under stress, behaves differently in microgravity than it would under Earth's gravitational pull.

By positioning LignoSat at the ISS, researchers could monitor subtle changes in the wood's structure, allowing them to understand how it might perform in microgravity. This data was valuable for refining wood treatments and structural designs, ensuring that future wooden satellites would maintain their integrity even in low-gravity environments.

A New Frontier in Sustainable Space Exploration

LignoSat's journey to the ISS was a landmark achievement in sustainable space exploration, showcasing the potential of biodegradable materials in a field traditionally dominated by metals and synthetics. The launch event, arrival at the ISS, and subsequent testing marked critical milestones, each contributing to our understanding of how eco-friendly materials can be used in space.

The strategic positioning of LignoSat at the ISS enabled comprehensive data collection on its response to radiation, temperature changes, and microgravity. The insights gained from this mission will be instrumental in guiding future efforts to develop sustainable satellites and space technologies that minimize environmental impact. As researchers analyze the data from LignoSat, they are paving the way for a future in which sustainable materials play a central role in space exploration, reminding us that innovation and environmental responsibility can, indeed, go hand in hand.

Through LignoSat, humanity is taking its first steps toward a greener space industry, one that values both technological progress and the preservation of our environment, both on Earth and beyond.

CHAPTER 5
Mission Objectives and Tests in Orbit

The deployment of LignoSat, the world's first wooden satellite, at the International Space Station (ISS) represents a major milestone in the field of sustainable space exploration. With LignoSat, scientists and engineers are seeking to answer a pivotal question: can wood, a traditionally Earth-bound material, survive and function in the harsh environment of space? The mission objectives of LignoSat are rooted in a six-month testing period during which researchers will observe and analyze the wood's performance. This chapter explores the key tests and experiments planned for LignoSat, describes how scientists will monitor the wood's response to space conditions, and explains how these findings could guide future innovations in sustainable construction for space exploration.

Key Tests and Experiments: The Six-Month Testing Period

The six-month testing period for LignoSat was designed with specific objectives to evaluate the performance of magnolia wood under real space conditions. This period involves a series of meticulously planned tests, each focusing on a different aspect of the satellite's materials, structure, and functionality. Scientists and engineers from Kyoto University and Sumitomo Forestry have collaborated with the ISS and other space agencies to ensure these tests cover all necessary conditions that wooden materials might encounter in orbit.

The primary tests are divided into key areas: thermal resistance, structural integrity, radiation exposure, vacuum outgassing, impact resistance, and microgravity effects. Each of these tests is essential for determining whether wood can endure space conditions over an extended period. The experiments have been structured to collect data progressively, allowing researchers to observe any gradual changes or cumulative effects on the wood.

1. Thermal Resistance Testing

Thermal resistance testing is one of the core objectives of LignoSat's mission. As a natural material, wood has inherent thermal insulation properties, but the extreme temperatures in space require it to withstand sudden shifts. While the temperature on the sunlit side of an object in low Earth orbit can rise to over 250°F, the shaded side can drop to -250°F. This constant cycling occurs every 90 minutes as the satellite orbits Earth, making thermal resilience crucial for any material in space.

The thermal resistance tests involve monitoring how magnolia wood reacts to these frequent temperature shifts. Sensors embedded in the wood structure continuously record temperature changes, tracking any signs of expansion, contraction, or warping. The data from these sensors are transmitted to Earth for analysis, allowing researchers to evaluate whether the wood's structural integrity is compromised by thermal stress. If successful, these tests could indicate that treated wood can serve as an effective thermal insulator in space, providing essential data for future material selection in sustainable satellite design.

2. Structural Integrity and Durability Testing

Another crucial objective for LignoSat is to assess the wood's structural integrity over time. In space, materials are subjected to stresses they would never encounter on Earth. For wood, these stresses include exposure to vacuum, microgravity, and space debris. Structural integrity testing focuses on observing whether the wood retains its shape, strength, and cohesion under prolonged space exposure.

Throughout the six-month period, LignoSat's structure is monitored for signs of cracking, splitting, or deformation.

Specially designed sensors measure tiny changes in the wood's surface and internal structure, helping scientists understand how it responds to stress. Durability tests also include monitoring the effects of micro-vibrations and minor impacts from micrometeoroids. Data from these tests will inform whether wood could serve as a reliable structural component in satellites, possibly reducing the reliance on metals and other resource-intensive materials.

3. Radiation Exposure Analysis

One of the most challenging aspects of space is the high level of cosmic radiation that materials are exposed to. Radiation can gradually degrade materials, especially organic ones like wood, affecting their structural properties and performance. The radiation exposure analysis focuses on how magnolia wood withstands radiation over time, with sensors recording radiation levels and detecting any changes in the wood's molecular structure.

For this test, samples of the treated magnolia wood are placed on different sides of the satellite to expose them to varying levels of radiation. Over the six-month period, researchers will analyze whether the wood begins to weaken, lose mass, or exhibit any other signs of degradation. Understanding how wood responds to radiation will be crucial for determining its viability as a satellite material, as well as for refining wood treatments to enhance radiation resistance.

4. Vacuum Outgassing Testing

The vacuum of space presents another unique challenge. In a vacuum, materials often undergo a process called outgassing, where trapped gases escape from the material. This can weaken the structure of the material and contaminate surrounding instruments. Wood, being a porous material, is particularly prone to outgassing, so it must be treated to minimize gas release.

In this test, researchers are monitoring the levels of gases released by LignoSat's wood components. Spectrometers on board the satellite measure any emissions, allowing scientists to track outgassing levels over time. This data is crucial for evaluating the effectiveness of the treatments applied to the wood before launch. If the treatments are effective, outgassing will be minimal, confirming that wood can be a viable option for satellite construction without compromising nearby instruments.

5. Impact Resistance and Micrometeoroid Testing

Space is filled with tiny particles, including micrometeoroids, that travel at high speeds. Even small impacts can cause damage to a satellite's structure, so materials must be able to withstand minor collisions. For LignoSat, impact resistance testing aims to measure how well wood can endure such impacts without sustaining severe damage.

Sensors embedded in the wood detect and record impacts, while high-speed cameras document any changes in the wood's surface. The data from these tests will indicate whether wood has the resilience needed to function in a space environment where minor collisions are inevitable. If successful, this data could support the use of wood as a durable material in space, potentially inspiring future innovations in impact-resistant treatments for wooden structures.

6. Microgravity Effects on Wood's Cellular Structure

One of the lesser-known effects of space is the influence of microgravity on materials. Microgravity can alter the way materials behave at a cellular level, sometimes causing unexpected shifts in structure or cohesion. LignoSat's mission includes studying how microgravity affects the wood's cellular arrangement and overall structure.

Using microscopic analysis, researchers can examine whether the cells within the wood undergo any significant changes due to the absence of gravity. This data is important for understanding whether wood's cellular structure can maintain its integrity in long-term space missions. If successful, the results could indicate that wood has a stable, predictable response to microgravity, supporting its use in future satellites and space-based constructions.

Monitoring Wood's Response to Space Conditions

The success of LignoSat's mission relies heavily on accurate, real-time monitoring of how magnolia wood reacts to space conditions. Each test outlined above requires constant observation to ensure that scientists capture even the smallest changes in the wood's performance. To accomplish this, LignoSat was equipped with a range of sensors and instruments that measure various environmental parameters and the wood's response to them.

Thermal Sensors: Tracking Temperature Changes

Thermal sensors were strategically embedded in the wood to track how it responds to temperature fluctuations. These sensors capture data on expansion, contraction, and any thermal stress fractures that may occur. Since temperature changes happen quickly in space, the sensors are designed to record data at high frequencies, providing a comprehensive profile of the wood's thermal behavior.

By transmitting data back to Earth in real time, these sensors allow researchers to monitor any signs of thermal fatigue or degradation. This information is critical for understanding whether wood can provide the thermal stability needed for space structures, possibly reducing reliance on traditional insulation materials.

Structural Sensors: Detecting Deformation and Stress

Structural sensors in LignoSat measure shifts in the wood's physical shape, such as bending, warping, or cracking. These sensors use tiny strain gauges to detect even the slightest changes in the wood's structure. If the wood shows signs of stress or deformation, the data will reveal whether these changes are temporary or permanent, helping scientists assess the wood's durability.

Radiation Detectors: Assessing Radiation Impact

Radiation detectors were installed to monitor the levels of cosmic radiation impacting the wood and to track any cumulative effects on its molecular structure. These detectors record the radiation levels at different points on the satellite, capturing data on both high-intensity radiation and background radiation levels. This data will help researchers understand the limits of wood's radiation tolerance and improve treatments for enhancing its resilience.

Outgassing Monitors: Measuring Gas Emissions

Outgassing monitors are responsible for measuring any gases released by the wood. These monitors use spectrometric technology to detect and analyze emissions, helping researchers determine whether the treatments applied to the wood are effective. By minimizing outgassing, scientists hope to make wood a viable, contamination-free material for satellite construction.

Impact Detectors: Recording Micrometeoroid Collisions

Impact detectors measure the force and frequency of micrometeoroid collisions on LignoSat. These detectors are designed to capture real-time data on the effects of high-speed impacts, providing insights into the wood's resilience. This information will inform future designs for impact-resistant wood treatments, making it possible to use wood in environments where collisions are inevitable.

Collecting Data for Future Wooden Structures in Space

The data collected from LignoSat's tests are invaluable for the future of sustainable space exploration. By understanding how wood responds to the challenges of space, scientists can begin to envision a future where biodegradable materials play a prominent role in satellite construction and even in building larger structures for human habitation.

Paving the Way for Sustainable Satellite Design

If the results of LignoSat's tests are promising, they could lead to a new generation of satellites built with eco-friendly materials. By using wood and other biodegradable substances, the space industry can reduce the environmental impact of satellites at the end of their operational lives. Wooden satellites would burn up in Earth's atmosphere without leaving harmful debris, addressing the growing issue of space junk and ensuring a cleaner orbital environment.

Expanding Material Options for Spacecraft Construction

The success of LignoSat could inspire researchers to explore other unconventional materials for space applications.

Wood's potential as a sustainable and resilient material may encourage scientists to consider additional organic compounds, such as biodegradable polymers or natural fibers, as alternatives to traditional metals. This diversification in material selection could lead to innovative spacecraft designs that prioritize environmental responsibility alongside functionality.

Laying the Foundation for Space Habitats

Beyond satellites, the findings from LignoSat could impact the construction of future space habitats. If wood proves durable and reliable in space, it could serve as a building material for modular structures on space stations or even lunar and Martian habitats. Using wood as a construction material in these contexts would reduce the need for heavy, resource-intensive materials, allowing for a more sustainable approach to off-Earth living spaces.

Advancing Knowledge in Material Science

LignoSat's mission will contribute significantly to the field of material science by providing valuable insights into how organic materials behave under space conditions. The knowledge gained will aid scientists in developing new treatments, coatings, and preservation methods for wood and other biodegradable materials, broadening the scope of materials suitable for space exploration. These advancements could lead to practical applications on Earth as well, promoting sustainability in industries that benefit from resilient, eco-friendly materials.

LignoSat's Legacy in Space Exploration

The mission objectives and testing phase of LignoSat represent a pioneering effort to bring sustainability to the forefront of space exploration. By testing wood in the most extreme environment imaginable, scientists are challenging traditional assumptions about material selection in satellite construction. The data collected from LignoSat's mission will inform future satellite designs, opening the door to a new era of eco-friendly space technology.

LignoSat is more than just a wooden satellite; it's a symbol of humanity's commitment to finding innovative solutions that respect our planet and extend our stewardship to outer space. Through this mission, researchers are proving that sustainability and technological advancement can coexist, setting the stage for a future in which space exploration honors the natural world while pushing the boundaries of human achievement. The journey of LignoSat is just beginning, but its legacy will inspire generations of scientists and engineers to reach for the stars in a way that is both responsible and visionary.

CHAPTER 6

Potential Impacts on Satellite Technology and Sustainability

The launch of LignoSat, the world's first wooden satellite, marks a significant shift in how we approach satellite technology and sustainability. By pushing the boundaries of conventional satellite materials and using wood, LignoSat opens new possibilities for reducing the environmental impact of space exploration. This chapter explores the potential impacts of wooden satellites on environmental sustainability, the broader future of eco-friendly materials in space, and how LignoSat's mission could catalyze change in the satellite industry, inspiring a new era of environmentally-conscious design and innovation.

Environmental Impact of Wooden Satellites: Reducing Space Debris

One of the most pressing issues in modern space exploration is the accumulation of space debris, which poses significant risks to both active satellites and future missions. Space debris, or orbital debris, consists of defunct satellites, rocket stages, and fragments from collisions or disintegrated equipment. These objects, often left in orbit for decades or even centuries, continue to travel at high speeds, presenting collision hazards to new satellites, the International Space Station (ISS), and other space-based assets. This growing debris field could eventually limit our ability to safely launch and operate satellites, making it a serious environmental and operational concern.

Wooden satellites like LignoSat offer a potential solution to this problem by being fully biodegradable upon re-entry. Traditional satellites are constructed from materials such as aluminum, titanium, and other composites that can survive re-entry and either contribute to space debris or fall back to Earth in the form of dangerous fragments. Wooden satellites, on the other hand, are designed to burn up completely upon re-entering Earth's atmosphere, leaving little to no trace behind. This characteristic can play a vital role in reducing the accumulation of space debris, making wooden satellites an attractive option for the future of environmentally-responsible space missions.

Complete Burn-Up During Re-Entry

Wooden satellites like LignoSat are designed to fully disintegrate upon re-entry, thanks to the organic nature of their primary material. Wood, unlike metals and composites, combusts at relatively low temperatures, making it suitable for complete incineration when it re-enters the atmosphere. By burning up fully, wooden satellites eliminate the risk of contributing to space debris or posing a hazard to Earth's surface.

When a wooden satellite re-enters the atmosphere, it begins to experience intense frictional heating, causing the wood to combust. This combustion process is quick and efficient, ensuring that the satellite is destroyed before it reaches lower altitudes. As a result, wooden satellites can complete their missions without leaving long-lasting debris, reducing the environmental footprint of satellite technology and setting a new standard for sustainable satellite design.

Addressing the Kessler Syndrome

The Kessler Syndrome is a theoretical scenario in which space debris reaches a critical density, causing a chain reaction of collisions that generates even more debris. This could make certain orbits unusable, endangering existing satellites and impeding future launches. By choosing biodegradable materials that leave no debris, wooden satellites like LignoSat could help mitigate the risks associated with the Kessler Syndrome.

If more satellites are designed with biodegradable materials, the amount of permanent debris in orbit could be significantly reduced. This could help ensure that future generations can continue to access and utilize space without the growing threat of uncontrollable debris. In this way, LignoSat's innovative use of wood serves as a preventive measure against the worst-case scenarios of space debris accumulation.

Minimizing Environmental Impact on Earth and Beyond

The environmental benefits of wooden satellites extend beyond reducing space debris. The production of wood is far less resource-intensive compared to metals and composites traditionally used in satellite construction. Growing wood for satellite use can be part of a renewable cycle, reducing the need for mining and metal processing, which are energy-intensive and have adverse environmental impacts. The sustainable cultivation of wood contributes to a lower carbon footprint, making wooden satellites an environmentally-friendly choice from production to re-entry.

Moreover, the use of wood aligns with global initiatives to minimize the ecological footprint of human activities, including those that extend beyond Earth. By using a biodegradable material, LignoSat demonstrates that space exploration does not have to come at the expense of environmental responsibility. It paves the way for future satellite designs that prioritize both functionality and sustainability, offering a blueprint for an era of eco-conscious satellite technology.

Future of Sustainable Materials in Space

While wood's successful deployment in LignoSat represents a pioneering step in sustainable satellite technology, it also raises an important question: What other materials could we use to reduce the environmental impact of space exploration? The future of sustainable materials in space goes beyond wood, as scientists and engineers look for new, innovative materials that balance durability, functionality, and biodegradability.

Biodegradable Polymers

One promising avenue for sustainable satellite construction is the development of biodegradable polymers. These materials, derived from renewable sources like cornstarch, cellulose, and plant-based oils, offer an alternative to traditional synthetic polymers, which can persist in the environment for long periods. Biodegradable polymers can degrade in specific conditions, such as when exposed to sunlight, heat, or specific enzymes. In a satellite context, biodegradable polymers could be used to construct lightweight components that either break down during re-entry or degrade over time if left in orbit.

Researchers are currently exploring ways to enhance the strength and resilience of these polymers to make them suitable for space applications. If successful, biodegradable polymers could be incorporated into satellite components such as casings, structural supports, and insulation layers, reducing the environmental impact of satellites once they reach the end of their operational lives.

Natural Fibers and Composites

Natural fibers such as hemp, flax, and jute, as well as composites made from these materials, are being investigated for their potential use in satellite construction. These fibers are renewable, biodegradable, and have favorable strength-to-weight ratios, making them suitable for lightweight satellite components. When combined with biodegradable polymers, natural fibers can form composites that are both strong and sustainable, potentially replacing conventional materials in non-critical satellite parts.

Natural fiber composites can also be engineered to have specific properties, such as improved thermal insulation or impact resistance. These properties could make them suitable for a range of satellite applications, including shielding, support structures, and interior components. By exploring natural fibers, researchers hope to create materials that not only perform well in space but also minimize environmental impact upon disposal.

Recycled and Recyclable Metals

Although metals are not biodegradable, there is growing interest in using recycled and recyclable metals to reduce the environmental impact of satellite construction. Recycled aluminum, for example, can be used in satellite casings and structural components without the environmental costs associated with mining and processing new metal. Additionally, designing satellites with recyclability in mind could enable the recovery and reuse of metal components, contributing to a circular economy in space exploration.

The concept of a circular economy in space involves designing satellites in a way that allows for easy disassembly and recycling of their parts. Future satellites could be designed with modular components that can be replaced, refurbished, or recycled, reducing waste and resource consumption. While this approach does not eliminate space debris, it reduces the need for new materials and promotes resource efficiency.

Nanomaterials and Self-Healing Materials

Nanotechnology offers a promising path for developing sustainable materials that are both lightweight and highly durable. Nanomaterials, such as carbon nanotubes and graphene, can enhance the strength and resilience of biodegradable materials, potentially making them viable for space applications. By incorporating nanomaterials into biodegradable polymers or natural fibers, researchers can create composites with exceptional performance while maintaining eco-friendliness.

Another exciting development is self-healing materials, which can repair minor damages caused by impacts or wear. These materials, still in the experimental phase, could extend the lifespan of satellites, reducing the need for replacements and minimizing waste. Self-healing materials could be especially valuable for long-term missions, where satellites may encounter micrometeoroids and other sources of damage. By repairing themselves, these materials could reduce the frequency of satellite replacement, contributing to a more sustainable approach to space exploration.

Changing the Satellite Industry: Inspiring Eco-Friendly Designs

The successful deployment of LignoSat has the potential to revolutionize the satellite industry, setting a new standard for eco-friendly design and inspiring other organizations to prioritize sustainability in their own satellite missions. While traditional satellite design has largely focused on durability and cost-efficiency, LignoSat demonstrates that environmentally-conscious design can be a viable and effective choice. This shift in priorities could have far-reaching implications for the satellite industry, encouraging a new generation of eco-friendly innovations.

Redefining Satellite Lifecycle Management

The concept of a satellite's lifecycle traditionally involves its design, launch, operational period, and eventual decommissioning. In most cases, decommissioned satellites become part of the ever-growing field of space debris. LignoSat's fully biodegradable design challenges this status quo by offering a lifecycle that minimizes environmental impact even at the end of the satellite's operational life.

As more companies and space agencies adopt the principles demonstrated by LignoSat, satellite lifecycle management could shift to include sustainability as a core consideration.

Future satellites may be designed with built-in deorbiting mechanisms or biodegradable materials, ensuring that they do not contribute to space debris. This approach aligns with global initiatives to reduce environmental impact and promote responsible resource use, paving the way for a satellite industry that values sustainability as much as performance.

Encouraging Corporate Responsibility in Space

The success of LignoSat is likely to inspire private companies, governments, and organizations to consider their environmental responsibility when designing and launching satellites. As public awareness grows about the importance of sustainability in space, there may be increased pressure on companies to adopt eco-friendly practices. LignoSat's mission can serve as a model for companies seeking to demonstrate their commitment to environmental responsibility, both on Earth and in orbit.

Corporate responsibility in space could also extend to regulations and standards for satellite design and decommissioning. International bodies may introduce guidelines or incentives for companies that use sustainable materials, reward eco-friendly designs, or require satellite operators to have deorbiting plans. By leading the way in environmental responsibility, LignoSat could contribute to a more regulated and conscientious approach to space exploration.

Expanding Innovation in Satellite Design

The introduction of wood as a satellite material highlights the potential for innovation in satellite design, pushing the boundaries of what materials and technologies can be used in space. As engineers continue to explore new materials inspired by LignoSat, they may develop designs that incorporate a mix of biodegradable, recyclable, and renewable resources. This could lead to hybrid satellite designs that combine traditional materials with eco-friendly alternatives, optimizing both performance and sustainability.

This expanded focus on material innovation could also result in the development of multi-purpose satellites capable of being reused or repurposed. For example, satellites could be designed with modular components that can be replaced or upgraded, reducing the need for new launches. By integrating flexibility and sustainability into satellite design, the industry could move toward a future in which satellites serve longer and more varied missions without adding to the problem of space debris.

Educating the Public and Inspiring Future Generations

The mission of LignoSat is not just a technical achievement but also an educational one. By demonstrating that sustainability and space exploration can go hand in hand, LignoSat has the potential to inspire future generations of engineers, scientists, and environmentalists. This project shows that space exploration does not have to come at the expense of environmental responsibility and that even the most advanced technologies can be aligned with sustainable values.

Educational initiatives that highlight the mission of LignoSat could encourage young people to consider careers in sustainable engineering and space sciences. By promoting a vision of space exploration that prioritizes environmental stewardship, LignoSat can help shape a new generation of scientists and engineers who are committed to responsible innovation.

Pioneering a Greener Future in Space Exploration

LignoSat's mission represents a groundbreaking step in the quest for sustainable space exploration. By proving that wood and other eco-friendly materials can be used in satellite construction, LignoSat has opened the door for a new era of environmentally-conscious satellite technology.

The environmental benefits of biodegradable satellites, the potential for alternative materials, and the influence on industry standards all point to a future where sustainability becomes a foundational principle in space exploration.

The success of LignoSat is a reminder that our actions in space have consequences that reach far beyond Earth. As we continue to push the boundaries of exploration, we must do so responsibly, ensuring that the technologies we deploy align with our commitment to protecting our environment. LignoSat is a beacon of that commitment, illustrating that innovation and environmental responsibility are not mutually exclusive but can coexist to create a greener, more sustainable future in space exploration. As the satellite industry continues to evolve, LignoSat's legacy will be a source of inspiration, reminding us that the future of space exploration is one where technology serves not just our ambitions, but the wellbeing of our planet as well.

CHAPTER 7

Expanding Applications of Wood in Space Exploration

As humanity ventures deeper into space and prepares for long-term missions, sustainable materials are becoming a priority in spacecraft design and construction. Wood, an ancient material with modern potential, is emerging as a viable option not only for satellites but also for future habitats and infrastructure on extraterrestrial bodies like the Moon and Mars. This chapter explores the possibilities of using wood in lunar and Martian habitats, examines the unique challenges and benefits of employing wood for long-term missions, and broadens the discussion to other experimental materials that hold promise for sustainable space exploration.

Wooden Structures on the Moon and Mars: Potential Uses in Extraterrestrial Habitats

With plans underway for lunar bases and, eventually, Martian settlements, architects, engineers, and material scientists are rethinking the materials that will form the backbone of these extraterrestrial habitats. Although traditional space infrastructure has relied on metal alloys, composites, and synthetics, the focus is now shifting toward sustainable, renewable materials that can be grown, processed, or even fabricated in space. Wood, with its durability, versatility, and biodegradability, presents a surprising yet intriguing option for building materials on the Moon and Mars.

The Case for Wood in Lunar Habitats

The Moon is expected to be a testing ground for long-term human presence beyond Earth, and establishing a base on its surface will require extensive infrastructure. Initial habitats will likely be modules transported from Earth, but long-term structures will need to rely on sustainable and locally sourced materials, both to reduce transportation costs and to facilitate maintenance and expansion. Wood, if adapted properly, could play a role in constructing these habitats, particularly for interior structures where exposure to the harsh lunar environment can be minimized.

Wooden materials could be used in several ways in lunar habitats:

- **Interior Panels and Partitioning:** For lunar bases, wood could serve as a lightweight and insulating material for interior panels and partitions. Treated wood can provide thermal insulation, reducing the need for extensive energy systems to maintain livable conditions within lunar habitats. The relative flexibility and workability of wood make it easy to install, repair, or reconfigure, which is valuable in confined environments like lunar modules.

- **Furniture and Storage Solutions:** Wood's aesthetic appeal and tactile qualities make it an excellent choice for furniture and storage within habitats. While furniture may seem secondary, creating comfortable, human-centered spaces is essential for the psychological wellbeing of astronauts on long-term missions.

- **Shielding and Insulation:** In the interior spaces of lunar habitats, wood's thermal and sound insulation properties could be advantageous. When paired with other materials, wood could help shield habitats from temperature extremes and radiation in lower-risk zones, such as storage areas or recreational spaces.

Potential Applications on Mars

Mars, with its thin atmosphere and relatively mild temperatures compared to the vacuum of the Moon, may offer more opportunities for using wood as a structural material. The possibility of creating an ecosystem that could support plant growth also opens doors for renewable material production on Mars, including growing wood-like fibers or creating composite materials with organic components.

Potential applications of wood on Mars include:

- **Structural Components for Pressurized Habitats:** Mars habitats must be pressurized to maintain livable conditions. Wood, when treated and protected, could form part of the structural framework for certain areas of these habitats. With local resources, there's potential to grow or produce bio composite materials that incorporate wood fibers, reducing the need to import construction materials from Earth.

- **Biodegradable Modules for Experimental Research Stations:** For scientific outposts that require periodic relocation, wooden structures could serve as biodegradable research modules. Once their operational life ends, these structures could be left to degrade or be composted as the mission expands, reducing the ecological footprint of human presence.

- **Hydroponic Farming and Biodome Enclosures:** Wood could be used to construct biodomes or hydroponic stations where crops are grown to sustain astronauts. The relatively mild Martian environment would allow for controlled conditions that support plant growth. Wood could act as a non-reactive material in contact with plants, and it could be locally sourced if bioreactors are established to produce wood fibers from Martian-grown plants.

Challenges and Considerations for Wood in Long-Term Missions

While wood offers exciting possibilities for space applications, it also presents unique challenges that must be addressed to ensure its viability in extraterrestrial environments. From structural resilience to potential environmental risks, these challenges highlight both the limitations and the potential of using wood for long-term missions on the Moon, Mars, and beyond.

Limitations of Wood in Space

1. **Vulnerability to Extreme Environments**
   Wood's organic nature makes it vulnerable to temperature extremes, radiation, and the lack of atmospheric pressure, all of which are characteristic of space environments. Wood, when exposed directly to these conditions, could degrade or fail over time,

impacting the structural integrity of habitats or equipment. To overcome these limitations, wood must be heavily treated, coated, or shielded in extraterrestrial environments. However, treatments may add weight or complexity to wooden structures, reducing some of wood's environmental benefits.

2. **Flammability**

Fire safety is a significant concern in space missions, as any material that could potentially ignite is a hazard. Wood is a combustible material, which poses risks in a pressurized environment with limited escape options. Treatments exist to make wood fire-resistant, but these treatments may lose effectiveness over time or add to the structural load. Developing non-flammable or fire-resistant versions of wood for space will be essential before it can be widely adopted.

3. **Decomposition and Pest Control**

In terrestrial settings, wood is susceptible to decomposition from fungi, bacteria, and insects. While space habitats lack these decomposers, the potential introduction of microbial life for food production or waste recycling could introduce risks of wood degradation. In closed environments, treating wood against microbial decay while ensuring treatments are non-toxic to inhabitants is crucial.

4. **Material Sourcing and Transport Costs**

    Although wood has promising applications, its availability in space is limited. Shipping wood from Earth would be costly, so sustainable sources on the Moon or Mars would be necessary for widespread use. Bioreactors or Martian greenhouses might be able to produce wood fibers in the future, but the technology is currently under development and may require years of refinement.

Benefits of Wood for Space Applications

1. **Thermal and Acoustic Insulation**

    Wood has natural insulating properties that could improve thermal management in habitats, potentially reducing energy requirements. Wood's porous structure provides sound-dampening capabilities as well, which can be useful in reducing noise within confined living quarters, enhancing astronauts' comfort and wellbeing.

2. **Renewable and Biodegradable**

   Wood's renewable nature aligns with sustainable resource cycles, particularly if it can be grown on Mars or synthesized in bioreactors. Its biodegradability could allow for a cyclical resource chain where materials are reused or composted after use. This advantage is beneficial for missions seeking to maintain minimal environmental impact.

3. **Psychological Comfort and Aesthetic Value**

   Wood is familiar, natural, and visually pleasing, which can have psychological benefits for astronauts living far from Earth. The presence of wood, with its warmth and texture, could improve morale and create a more home-like environment. This emotional benefit is significant for the mental health of astronauts on long-duration missions.

4. **Material Versatility and Lightweight Properties**
   Wood's versatility allows it to be easily shaped, joined, and repaired. Its lightweight nature can also reduce launch costs, though this advantage depends on the density and type of wood used. In particular, wood composites may offer an ideal balance of

weight and strength, further enhancing wood's appeal in space applications.

Other Experimental Materials for Spacecrafts: The Broader Context of Sustainable Materials

Wood is only one of many materials being considered for sustainable space missions. Researchers are exploring various eco-friendly, renewable, and even synthetic materials that could reduce environmental impact and improve the safety, functionality, and efficiency of space structures. This section provides an overview of other experimental materials that may complement or compete with wood in future spacecraft and habitat designs.

Biopolymer Composites

Biopolymer composites, made from renewable sources such as starches, proteins, and cellulose, are increasingly used in terrestrial applications and show potential for space missions. These materials are biodegradable, can be strengthened with additives like carbon fibers or nanomaterials, and could serve as alternatives to traditional composites. Biopolymer composites could be ideal for creating lightweight panels, insulation, and interior furnishings in space habitats, where biodegradability and low environmental impact are critical.

Mycelium-Based Materials

Mycelium, the root structure of fungi, is being explored as a sustainable material that could be grown into structural components. Mycelium can form durable, lightweight, and biodegradable materials that are suitable for insulation and structural support. NASA has experimented with mycelium as a potential material for Martian habitats, as it could be grown on-site in controlled environments. Mycelium's insulating properties and ability to grow into specific shapes make it an attractive option for sustainable habitat construction.

Synthetic Spider Silk and Other Protein Fibers

Synthetic spider silk, a bioengineered material that mimics the properties of natural spider silk, has a high tensile strength and flexibility. This material, along with other bioengineered protein fibers, is being explored for its potential in creating strong, lightweight, and biodegradable structures for space. Synthetic spider silk could be woven into fabrics for habitats, cables, or repair materials, offering resilience and versatility in long-term missions.

Martian Concrete and In-Situ Resource Utilization (ISRU)

In-situ resource utilization (ISRU) refers to the use of local resources on the Moon, Mars, or other celestial bodies to produce building materials. Martian concrete, made from a combination of Martian soil, water, and binding agents, has been proposed as a potential building material for habitats on Mars. This form of concrete could be used for structural elements, providing a durable and locally sourced material that reduces reliance on Earth-based resources. ISRU has the potential to make space exploration more sustainable and cost-effective by leveraging local materials.

Self-Healing Polymers and Coatings

Self-healing materials, capable of repairing themselves after sustaining damage, have emerged as promising candidates for space applications. These materials can seal minor punctures, resist micro-meteor impacts, and potentially extend the lifespan of habitats and equipment. Self-healing polymers and coatings could be used in combination with wood or other materials, enhancing durability and reducing the need for extensive repairs. These materials are particularly valuable in space, where maintenance capabilities are limited.

Carbon-Based Nanomaterials

Carbon-based nanomaterials, such as carbon nanotubes and graphene, offer exceptional strength, thermal conductivity, and flexibility. These nanomaterials can reinforce other materials, including biopolymers or wood, to create composites with high-performance properties. Carbon nanomaterials could enable the construction of ultra-light, durable structures that withstand space conditions while maintaining a low environmental footprint. Their versatility and strength make them ideal for both structural and protective applications.

A Sustainable Future for Space Exploration

The exploration of wood and other sustainable materials in space represents a transformative shift in how we approach off-Earth construction and resource management. By considering materials like wood, biopolymers, mycelium, and synthetic fibers, scientists and engineers are laying the foundation for an era of space exploration that values sustainability, renewability, and environmental responsibility.

The use of wood in missions to the Moon and Mars could enable habitats that are not only functional and durable but also support closed-loop resource cycles, where materials are renewed and repurposed as needed. Wood, along with other biodegradable and locally sourced materials, offers a vision of space exploration that is harmonious with natural ecosystems, minimizing waste and maximizing the human potential to adapt to new environments.

As we move forward into this exciting chapter of human history, the challenge will be to integrate these materials in ways that honor the principles of safety, functionality, and sustainability. LignoSat and the pioneering efforts to test wood's viability in space represent the first step in this journey, proving that the future of space exploration need not be confined to metals and synthetics. Instead, it can be a future shaped by innovation, creativity, and an enduring respect for the natural world, both on Earth and in the stars.

CHAPTER 8

The Future of Eco-Friendly Space Exploration

As humanity ventures further into space, balancing exploration with environmental responsibility becomes a pressing concern. The launch of LignoSat, the world's first wooden satellite, represents a groundbreaking effort to integrate sustainability into satellite technology. With an emphasis on renewable and biodegradable materials, LignoSat embodies the concept of eco-friendly space exploration, aiming to reduce the environmental impact of space missions. This chapter delves into the implications of LignoSat's success or failure, the potential for sustainable materials like wood to transform space research, and the promising path forward for eco-friendly satellites in the context of LignoSat's journey.

Implications of LignoSat's Success or Failure

The mission of LignoSat is more than just a test of whether wood can withstand the conditions of space. It's a landmark in exploring the viability of sustainable materials in the harsh environment of low Earth orbit (LEO) and potentially beyond. The success or failure of LignoSat has far-reaching implications that could impact the trajectory of satellite design, mission planning, and sustainability standards in space exploration.

Success: A Paradigm Shift in Satellite Technology

If LignoSat proves to be successful, its results could signify a paradigm shift in how satellites are designed and built. A successful mission would demonstrate that wood, when properly treated and engineered, is capable of withstanding radiation, thermal fluctuations, and microgravity. This could open the door to a new class of biodegradable satellites, which would burn up upon re-entry and leave little to no debris, addressing one of the biggest environmental issues associated with space missions—orbital debris.

A successful LignoSat mission could:

- **Inspire Industry-Wide Adoption of Biodegradable Materials**

  Success would likely spur interest across the aerospace industry in exploring biodegradable materials for satellite construction. Companies and agencies might begin investing in research and development (R&D) focused on adapting sustainable materials for use in LEO and beyond. With biodegradable materials, satellites could be designed to leave no trace once their mission ends, helping to alleviate the growing problem of space junk.

- **Lead to New Standards in Spacecraft Design**
If LignoSat's mission results validate wood's resilience, regulatory bodies like the United Nations Committee on the Peaceful Uses of Outer Space (COPUOS) might establish guidelines or standards that promote biodegradable materials in satellite construction. This would represent a significant shift in spacecraft design philosophy, where environmental responsibility becomes a core criterion alongside performance and cost-efficiency.

- **Encourage Investment in Eco-Friendly Satellite Technology**
A successful outcome could encourage private and public investments in eco-friendly technologies, supporting further research into sustainable materials and manufacturing processes. Companies and governments may see LignoSat's success as a demonstration of the commercial viability of green

technologies in space, accelerating innovation in sustainable satellite technology.

Failure: Reassessing the Path to Sustainability in Space

On the other hand, if LignoSat encounters significant issues or fails to meet its objectives, the results would offer valuable lessons on the limitations of using natural materials in space. While failure would be disappointing, it wouldn't necessarily close the door on sustainable space exploration. Instead, it would highlight areas that require additional research, new materials, or innovative engineering approaches to make eco-friendly satellites feasible.

If LignoSat fails, it could lead to:

- **Improved Research on Material Treatments** Failure would provide insights into what aspects of wood need further modification to survive in space. Researchers could focus on advanced treatments, coatings, or hybrid designs that combine wood with other materials for enhanced durability. Even a partial failure could yield data essential for refining wood treatments, enhancing the chances of success for future missions.

- **Exploration of Alternative Biodegradable Materials**

  A failed mission might prompt scientists to explore other biodegradable materials with more inherent resilience to radiation, temperature extremes, and microgravity. For instance, biopolymer composites, natural fibers, or genetically engineered materials might emerge as viable options. These materials could be adapted to offer similar benefits to wood but with enhanced durability in space.

- **A Shift in Strategy Toward Hybrid Solutions**

  A failure may also encourage a shift toward hybrid solutions that combine sustainable and conventional materials, balancing eco-friendliness with resilience. Hybrid designs could pave the way for satellites that incorporate biodegradable components where possible, without compromising performance or safety.

Broadening the Scope of Sustainable Space Research

LignoSat has already expanded the conversation about sustainable materials in space, showing that renewable resources can be a viable option for spacecraft. However, LignoSat is just one step in the broader scope of sustainable space research, which includes a variety of materials and technologies aimed at reducing the environmental impact of human activities in orbit and beyond. As space missions become more frequent and ambitious, there is a need to explore and adopt practices that align with sustainability principles.

Expanding Research into Biodegradable and Renewable Materials

The success of LignoSat would bolster the case for extensive research into biodegradable and renewable materials that can endure space conditions. Potential areas of exploration include:

- **Biopolymers and Composites**

  Biopolymers, derived from renewable resources, can be engineered to degrade naturally or withstand specific environmental conditions. When combined with other materials, they can offer enhanced durability and functionality, potentially replacing synthetic polymers in spacecraft. For example, materials like polylactic acid (PLA) or polyhydroxyalkanoates (PHAs) can be customized for flexibility, strength, and even biodegradability.

- **Mycelium and Fungal-Based Materials**

  Mycelium, the root structure of fungi, is gaining attention as a sustainable material that could be cultivated and grown into specific shapes. With potential applications for both insulation and structural components, mycelium materials could offer an innovative approach to eco-friendly construction on extraterrestrial surfaces. Using mycelium and other fungal-based materials could further reduce reliance on Earth-sourced materials for space infrastructure.

- **In-Situ Resource Utilization (ISRU) for Sustainable Construction**

  As missions extend to the Moon and Mars, in-situ resource utilization (ISRU) will allow researchers to produce materials on-site rather than relying on resources from Earth. ISRU-based sustainable materials, potentially derived from Martian regolith or lunar dust, could be combined with biodegradable materials like wood or polymers. This approach would reduce transportation costs and facilitate sustainable infrastructure development on other planets.

Promoting Circular Economy Practices in Space

A successful LignoSat mission could contribute to a shift toward circular economy practices in space, where materials are reused, repurposed, and recycled to minimize waste. A circular approach could have profound benefits, especially for long-term missions and the eventual establishment of human settlements on other celestial bodies.

Key practices of a circular economy in space could include:

- **Reusable Components and Modular Satellites**
  Designing satellites with modular components that can be replaced or repurposed is a practical step toward sustainability. Parts made from biodegradable materials could serve temporary functions and be safely discarded, while reusable components could extend a satellite's operational life, reducing the need for new launches.

- **Recycling in Orbit**

  Recycling in space could be feasible if biodegradable materials like wood are incorporated into satellite designs. Systems that break down and repurpose materials in orbit would be valuable for missions requiring long-term sustainability. Such an approach could eventually support recycling stations in LEO, where outdated or decommissioned satellites are collected, dismantled, and repurposed.

- **Biodegradable Materials in Satellite Disposal**
  For satellites that cannot be recycled or repurposed, biodegradable materials could be used to ensure that they degrade harmlessly upon re-entry. This approach aligns with efforts to reduce space debris, preventing defunct satellites from cluttering orbit and threatening active spacecraft.

Setting Precedents for Sustainable Mission Planning

By showcasing eco-friendly materials, LignoSat could encourage space agencies and private companies to integrate sustainability as a core element of mission planning. This shift in perspective would impact every phase of mission planning, from design and development to deployment and decommissioning.

Agencies could incorporate sustainability criteria into their procurement processes, favoring contractors who prioritize environmentally responsible materials and methods. Private space companies might be incentivized to adopt eco-friendly practices, particularly as the demand for sustainable technologies increases on Earth and in space.

The Path Forward for Sustainable Satellites: Lessons from LignoSat and Future Possibilities

LignoSat's journey offers valuable insights and potential lessons for the future of eco-friendly space exploration. Whether the mission succeeds or encounters obstacles, it has already demonstrated that sustainability and space exploration can coexist. This project represents the beginning of a new era, where technological advancement and environmental responsibility go hand in hand.

Key Takeaways from LignoSat's Journey

1. **Innovation Is Essential to Sustainability in Space**
   LignoSat exemplifies how innovation drives sustainability. By using wood—a material seldom associated with high-tech applications—LignoSat challenges conventional thinking and pushes the boundaries of material science. The success of LignoSat could inspire more creativity in sustainable design, proving that traditional materials, when adapted, have a place in the future of space exploration.

2. **Sustainable Satellites Offer Practical Environmental Benefits**

The environmental benefits of biodegradable satellites are undeniable. By eliminating the risk of space debris and reducing the environmental costs associated with satellite construction and disposal, eco-friendly satellites offer a pragmatic solution to some of the space industry's most pressing environmental challenges. The mission of LignoSat could encourage more widespread adoption of sustainable satellite technology as stakeholders recognize its long-term value.

3. **Collaborative Efforts Are Crucial for Progress**
The success of LignoSat depends on collaborative efforts between research institutions, space agencies, and private companies. This project highlights the need for shared responsibility in advancing eco-friendly technologies. As space exploration becomes more accessible and competitive, collaborative projects will be essential to fostering innovation and setting sustainability standards that benefit the industry as a whole.

Future Possibilities for Eco-Friendly Space Exploration

Looking beyond LignoSat, there are numerous opportunities for expanding eco-friendly practices in space exploration. As technology advances and our knowledge of sustainable materials grows, the concept of eco-friendly space missions could become mainstream.

1. **The Development of Fully Biodegradable Satellites**

   With further advancements, it may be possible to construct fully biodegradable satellites that perform all necessary functions and degrade completely at the end of their lifecycle. These satellites would require minimal intervention for disposal, addressing space debris concerns and lowering operational costs.

2. **Eco-Friendly Space Stations and Lunar Bases**

   As humanity sets its sights on permanent settlements on the Moon and Mars, eco-friendly construction methods will be essential. Wood and other biodegradable materials could contribute to habitats that are designed to minimize waste, integrate renewable resources, and operate with circular economy principles. Lunar bases and Martian habitats built from sustainable materials would reflect a commitment to protecting the extraterrestrial environments we explore.

3. **Bioregenerative Life Support Systems**
The integration of biodegradable materials can extend to life support systems, where organic materials play a role in waste management, water recycling, and air purification. For instance, wooden structures could be part of closed-loop ecosystems, contributing to nutrient cycling in hydroponic systems or biodomes for plant cultivation.

4. **Next-Generation R&D on Hybrid Eco-Materials**
Future research may focus on hybrid materials that combine the strength and durability of conventional materials with the biodegradability of organic materials. Such advancements could lead to the creation of new composites specifically engineered for space applications, optimizing both performance and environmental impact.

5. **Policy and Regulatory Support for Sustainable Space Practices**

As eco-friendly technologies advance, regulatory bodies may play an active role in promoting sustainability in space. Policies that incentivize the use of biodegradable materials, regulate space debris, and prioritize environmental responsibility in mission planning would help establish industry-wide standards.

A Greener Vision for the Future of Space Exploration

The mission of LignoSat represents a pioneering step toward a more sustainable approach to space exploration. By integrating wood into satellite design, LignoSat challenges traditional practices and offers a blueprint for future eco-friendly space technologies. Whether the mission fully succeeds or encounters obstacles, the insights gained from LignoSat will inform and inspire the next generation of satellite technology, where sustainability becomes a foundational principle.

As we venture further into space, we carry the responsibility to minimize our environmental impact and adopt practices that honor the ecosystems of both Earth and the extraterrestrial environments we explore. The future of space exploration is not just about reaching new frontiers; it's about doing so in a way that respects and preserves the beauty of our universe. LignoSat has shown that sustainable space exploration is not only possible but necessary, lighting the path for future missions that balance the spirit of discovery with environmental stewardship.

CONCLUSION
The Wooden Satellite Handbook

As we reach the end of *The Wooden Satellite Handbook*, it's clear that the journey of creating, launching, and studying LignoSat—the world's first wooden satellite—marks a pivotal point in space exploration. This mission was born out of a desire to merge traditional materials with cutting-edge technology, forging a path towards more sustainable practices in an industry that often prioritises durability, cost, and efficiency over environmental responsibility. Yet, with the escalating urgency to address our impact on both Earth and space environments, LignoSat exemplifies how sustainable values can be integrated into even the most high-tech domains.

This concluding chapter reflects on the lessons learned from LignoSat's journey, the broader implications of using wood and other renewable materials in satellite design, and the future possibilities for sustainable space exploration. From LignoSat's development and launch to the challenges and innovations encountered along the way, each step has set a precedent for how the space industry might balance technological advancement with environmental stewardship.

Reimagining Materials in Space Exploration

Historically, materials used in space exploration have leaned towards the artificial and industrial. Metals, composites, and synthetics, though functional, are often resource-intensive and rarely designed with biodegradability or recyclability in mind. In contrast, LignoSat challenges the notion that these conventional materials are the only options for creating robust and reliable space technologies. By choosing wood as the primary construction material, scientists and engineers demonstrated that sustainability and durability can coexist, even in one of the most challenging environments known to humanity.

Wood's properties, such as its strength-to-weight ratio, natural insulating qualities, and biodegradability, make it a promising alternative to traditional satellite materials. Throughout this book, we've examined how magnolia wood was carefully selected and treated to endure space's harsh conditions. While wood does have limitations—vulnerability to radiation and temperature extremes being some of them—the success of LignoSat has shown that with the right adaptations, wood can survive and function effectively in space. This knowledge encourages a broader exploration of renewable materials for other space applications, such as habitat construction, equipment, and research facilities on the Moon and Mars.

The LignoSat mission has thus redefined the possibilities for materials in space, proving that traditional materials with sustainable characteristics can play a role in modern satellite technology. As space agencies and private companies look to the future, LignoSat's success encourages a more inclusive approach to material selection—one that considers the environmental impact of materials throughout their entire life cycle, from Earth to orbit and back.

Pioneering Environmental Responsibility in Orbit

One of LignoSat's most significant contributions to the field of space exploration is its emphasis on environmental responsibility. Space debris, or "space junk," is a growing concern for scientists and engineers, with defunct satellites, discarded rocket stages, and debris fragments accumulating in Earth's orbit at an alarming rate. These objects, travelling at tremendous speeds, pose a threat to operational satellites, future missions, and even human life on the International Space Station. The long-lasting nature of conventional satellite materials has exacerbated this problem, creating an orbital landfill that could hinder humanity's ability to safely explore space.

LignoSat addresses this issue by proposing a satellite that burns up completely upon re-entry. Wood, as an organic material, combusts fully and leaves little to no residual debris, minimizing its environmental footprint in space. This characteristic of wooden satellites introduces a sustainable solution to the problem of space debris, presenting a viable alternative to conventional satellites that often leave lasting remnants in orbit. If more satellites are designed to burn up cleanly, the risk of space debris collisions would be significantly reduced, helping maintain a safer and cleaner orbital environment for future generations.

LignoSat's success offers a new model for satellite lifecycle management, suggesting that biodegradability and environmental impact should be considered when designing and deploying new space technologies. This mission has demonstrated that satellites can be designed not only to perform and endure but also to return to Earth gracefully, leaving no lingering impact in the skies above.

Expanding the Role of Wood in Space Missions

The LignoSat project has highlighted wood's potential in satellite construction, but it also hints at a broader role for wood and similar materials in space missions. As humanity sets its sights on more ambitious goals—such as establishing a sustainable presence on the Moon or colonizing Mars—the need for lightweight, versatile, and renewable construction materials will only increase. Wood and other organic materials, if adapted properly, could serve as sustainable building blocks for extraterrestrial habitats, infrastructure, and even farming systems.

For instance, wood could be used to create insulated panels and partitioning in lunar habitats, offering thermal stability and soundproofing for astronauts. Similarly, wooden frames and composite materials could be incorporated into Martian greenhouses, providing a stable environment for crop growth. With advancements in bioreactors or other cultivation technologies, it may even be possible to grow wood-like fibers directly on Mars, reducing the need to transport materials from Earth and supporting a more self-sustaining settlement model.

The success of LignoSat is an important first step toward integrating wood and similar materials into a wider array of space applications. While challenges remain, the knowledge gained from this mission provides a foundation for expanding the role of wood beyond satellites, potentially enabling a sustainable future where renewable resources support human life on other celestial bodies.

A Model for Collaboration and Innovation

LignoSat exemplifies the power of collaboration and innovation in advancing sustainability in space exploration. The project brought together researchers, engineers, and scientists from Kyoto University, Sumitomo Forestry, and multiple space agencies, creating a synergy of expertise that allowed this ambitious vision to become a reality. By pooling resources and knowledge, these organizations achieved a milestone that would have been difficult to realize independently.

This model of collaboration could serve as a blueprint for future eco-friendly space missions, encouraging partnerships that bring together diverse perspectives and skill sets. As the space industry grows and becomes more competitive, collaborative projects like LignoSat underscore the value of shared goals and resources. The success of LignoSat is a testament to what can be achieved when entities work together towards a common vision, especially in the pursuit of responsible and sustainable exploration.

The mission also highlights the importance of innovation and adaptability in tackling environmental challenges. By rethinking what materials are suitable for space, the LignoSat team challenged conventional wisdom and pushed the boundaries of what was considered possible. This willingness to innovate, even with a material as seemingly simple as wood, inspires future scientists and engineers to consider unconventional solutions, fostering a culture of creativity in the pursuit of sustainability.

The Path Forward: Sustainable Satellites and Beyond

As we conclude *The Wooden Satellite Handbook*, it is worth considering the larger implications of LignoSat's journey and the potential future of sustainable satellites. Whether LignoSat's mission is a resounding success or encounters setbacks, it has already achieved its most important goal: sparking a conversation about the environmental impact of space exploration and demonstrating that sustainability can be a central tenet of satellite design.

Looking ahead, LignoSat's legacy could inspire new generations of scientists, engineers, and policymakers to prioritize sustainability as they develop and launch future space missions. As we seek to explore and understand our universe, we must also commit to protecting the environments we encounter and preserving the integrity of our home planet. The lessons from LignoSat offer a guidepost, showing us that even the sky is not the limit when it comes to environmental responsibility.

In the coming years, we may see a new class of eco-friendly satellites, designed with biodegradable materials, recyclable components, and low-impact disposal systems. Beyond satellites, sustainable practices could extend to lunar bases, Martian habitats, and interplanetary vessels, creating a future where space exploration aligns with the principles of stewardship and sustainability. By building on the foundation laid by LignoSat, humanity has the opportunity to forge a path that respects both the cosmos and the ecosystems we call home.

Final Reflections: A New Era of Responsible Exploration

The story of LignoSat reminds us that space exploration is not solely about reaching new destinations or advancing technology; it is also about making thoughtful choices and embracing our role as stewards of space. As the first wooden satellite, LignoSat is more than just a scientific experiment—it is a symbol of humanity's commitment to exploring responsibly and minimizing our impact on the universe.

LignoSat's journey challenges us to rethink our assumptions, expand our perspectives, and approach each new frontier with respect for the environments we encounter. It is a reminder that sustainability is not a constraint but a catalyst for innovation, inspiring us to develop technologies that benefit both current and future generations.

As we close this handbook, let us carry forward the lessons of LignoSat and envision a future where space exploration and sustainability are inseparable. In that future, each mission, satellite, and habitat we send into space will embody our commitment to leaving the universe a little better than we found it—cleaner, safer, and full of possibility for those who will one day follow in our footsteps. LignoSat has shown us that this vision is not only possible but achievable, inviting us to embrace a new era of responsible exploration, grounded in the values of creativity, collaboration, and environmental consciousness.